Interviews With My Godfather

Interviews with Entomologist Dr. Edward L. Mockford

Interviews Conducted by Juventino Manzano

Dedicated, of course, to my Godfather.

ISBN# 978-1-312-48306-4

CONTENTS

Videos available at:

https://www.youtube.com/watch?v=Q93VXBIIX2Uw&t=396s

https://www.youtube.com/watch?v=9asWeaYNSvA

The Discovery of the World's Smallest Insect

Interview conducted with Dr. Edward Mockford on May 4, 2015.

Today I am talking to my Godfather, a very important entomologist, Dr. Edward Mockford, and what we're going to talk about today is the discovery of the world's smallest insect. Doctor, would you tell us a little bit about yourself?

Yes, I've been studying a group of insects called bark lice or *psocids* for many years. I began studying them when I was fifteen, and I've just never dropped them and we're in more different times so there's a bigger and bigger backlog. Some years ago, in about 1998 or 99, I got interested in the life history of one of these bark lice, which is a species called *Echmepteryx hageni.* It is a little scaly winged creature; it looks like a very, very, small moth-- scaley wings just like a moth, and it's found on tree trunks and branches all throughout Eastern North America up into Southeastern Canada. This creature is of interest in a number of ways; most of its populations are parthenogenetic, that is to say, they reproduce without sex. All females, and I was very interested in the fact that it forms two kinds of eggs; it forms a summer egg of which the early generations of the of the year make, which turns orange, bright orange, before it hatches and stays bright orange. Then later in the season, it forms a different kind of egg, a winter egg, which turns black, and that will survive the winter, more or less, no matter how long or how rough the winter is. These eggs are laid under the bark of a tree, so they'll make it through winter. So, I was just really discovering these facts about the egg, which in fact have never been published yet, and I kept some wintering eggs on my screened in back porch in Central Illinois. Of course, they were going to be exposed to winter temperatures and I was going to see, okay, do these black eggs really make it through the winter?

About February, I think this was about February of the year 1999, I decided okay, I'm going to open one or two of these eggs and try to determine what stage the embryos are in.

Well, I opened one egg and to my surprise I did not find a bark louse embryo but instead, I found a fully developed female of a little parasitic wasp of the family *Mymaridae*, the so-called fairy flies; they're actually, little wasps, not flies, but that's their common name, fairy flies, because of their delicate looking bodies. Well, here was this female; she took up nearly the whole egg, so she was about half a millimeter in length. In addition, I found three in the same egg; three tiny little creatures that looked almost like mites. I was very lucky to be able to get those very minute creatures on to microscope slides, into a mounting medium and was then able to observe them carefully. It was very obvious that they were not mites with the usual body parts--six legs, and it was also, very obvious that they were males: they had, sizable male genitalia showing in the abdomen—well, what on earth were they? I knew a little bit about fairy flies, not very much, they were certainly way out of my specialty, but I knew that some of them at any rate, had males and females that looked very much alike.

So, were these little creatures the males or the female that I found in the same egg? Well, they must have been on careful examination, though they looked very different. The females had an antenna that was essentially two segments with one big bulgy segment out on the end of a very small basal segment and that this big bulky thing made it look like the head had two ears sticking out from the head and as a very wise specialist later joked with me about it, this looked like Mickey Mouse, but they were quite real creatures and I was very puzzled. Then, I recalled the life history of the common fig wasp which has a wingless male. I decided, okay, this must fit into the smaller sort with some of which at any rate, have wingless males. That's not the only curiosity about these creatures, which was the wingless males, but their very small size and also the fact that they had suction cups on the ends of their feet. As far as I was aware, there was no form in the literature that had suction cups on the ends of its feet. So, I got in touch with the specialist on fairy flies at the US National Museum and also, the specialist at the Canadian National Museum. The latter gentleman was very helpful in getting me on to the necessary literature to deal with this little creature. Turns out that my find was a member of a genus called *dycopomorpha* and that there were no forms of *dycopomorpha* known in North America that had any males. I think there was perhaps one species known in North America, but it didn't have males and, in some respects, it differed from the one I had. The other known species of *dycopomorpha* were from Argentina so, this was a brand-new form and I decided okay, I'm going to go ahead and describe this creature. Working in a different group of insects is a little bit like traveling to another universe: all is different. You would think that most of the anatomy is the same, but the fact is that all of the anatomical details are different, and you have to do a lot of learning of new information before you can attack the description of a new species. So, that's what I did for a number of months and got what I thought was a pretty good description together. I sent it to Dr. Huber, John Huber, at the Canadian National Collection, and he very kindly went over my paper in great detail and pointed out a number of things. So, I got the paper into close to publishable shape and I decided okay, this one should go to a journal that has a good readership, and I sent it to *the Annuals of the Entomological Society of America*. We were rather limited in the number of pages within; that they have free, at any rate, in that journal, but I was able to get the paper small enough, so that I was able to publish it at no cost, since I am a member of that organization. The paper came out in, I think it was the first issue of *the Annuals of the Entomological Society.*

In '01, I did not make the claim that these males were the smallest insect. I knew that they must be close to the smallest size that an adult insect can take because it's just impossible to imagine an adult insect functioning in a smaller body than that. Dr. Huber assured me that this is the smallest known adult insect, and so, I believe I did mention that Huber, who is an expert on these very tiny wasps, *Mymaridae*, had made this assurance to me and that it seems quite likely that this *is* the smallest known insect. Well, here we are in 2015 and nobody has actually, challenged this as being the smallest known insect. There is, I did see somewhere, a paper by someone, I think it was in India, or about specimens from India which apparently are of about the same size as my *dycopomorpha*, which I named by the way, as for the host species *echmepterygis*, so I called my new species *dycopomorpha echmepterygis*, ending in a possessive form. It seems like a very long name for a very small insect, but I felt that that was the name necessary. So, are there any other questions?

Wow. It's really amazing that you're able to remember in such precise detail every single step of this discovery. That's just amazing. I'm sitting here like, my goodness. It's like listening to an artist describe the process of creating their work. So, this has never been put out into the popular culture. This has been something that's been kind of staying in the field of entomology?

A Japanese person contacted me a number of years ago and said that they were at some museum in Japan, I've forgotten which, and that they had read my paper and could I possibly send them some specimens that they wanted to make an exhibit with in a museum? So, I did send them a couple of specimens that I still had on slides. Incidentally, the types of this species all went to the U.S. National Museum and I also sent specimens to the Canadian National Museum and also, to one of the insect collections in California. I think it was the University of California at Riverside, so they all had representatives of this, including the small male.

That's amazing. So, you describe him as, it, him, looking like Mickey Mouse and there's been nothing else that's been mentioned other than these Indian specimens. There's been nothing else that's in the similar category that's that small, that's it.

That's it right, there are very few insects that can function at that small a size. Unfortunately, there was an invertebrate physiologist at our university who wanted to try to get some sort of an estimate on the number of cells in the body of this creature, but he never did manage to do it, and I wish he had because it would be an interesting thing to pursue, but it never happened.

Well, thank you very much for this very enlightening interview, Dr. I appreciate it very much, and I'm sure that there will be many people out there that'll be interested in this story.

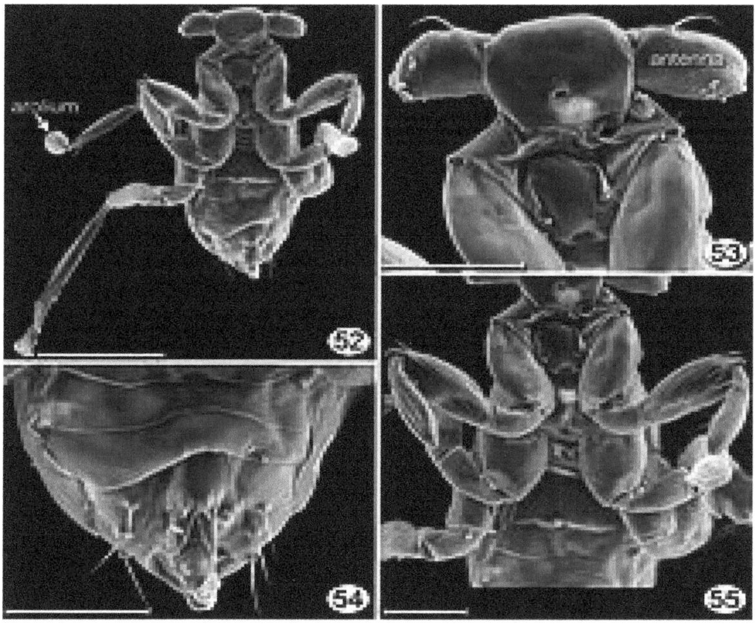

Defining the Order. This vast assemblage of insects is second only to *Coleoptera* (Beetles) in the number of described species. Of the 6,000–7,000 new species of insects described annually, *Hymenoptera* is a large component, especially in the parasitic wasp groups. Nearly all commonly encountered *Hymenoptera* can be recognized by a narrow "waist." When winged, the wings form two membranous pairs that can be hooked together. Ovipositors of *Hymenopteta* are usually well developed and modified into a stinger in the higher forms of the order. Because the "stinger" of such forms has developed from the ovipositor of females, male wasps are not able to sting. Many species of *Hymenoptera* are extremely small and are thus difficult to identify even to family. A publication by Edward Mockford in 1997 recorded the discovery of a new species of tiny wasp that is now known as the tiniest existing insect.

https://www.si.edu/spotlight/buginfo/hymenoptera

More information about Dr. Mockford's research:

https://www.researchgate.net/scientific-contributions/Edward-L-Mockford-2002043070

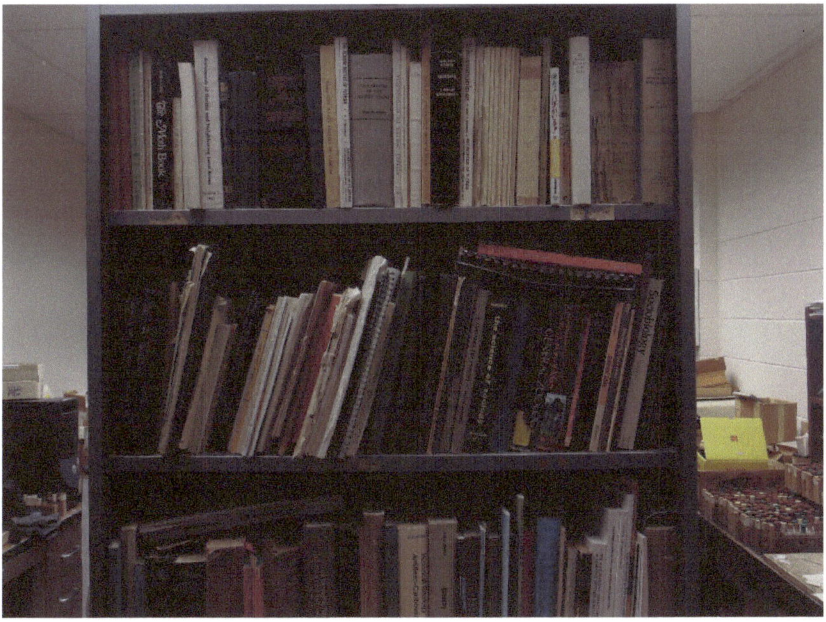

A bookshelf in his lab at ISU (above). Dr. Mockford on his microscope in his lab.

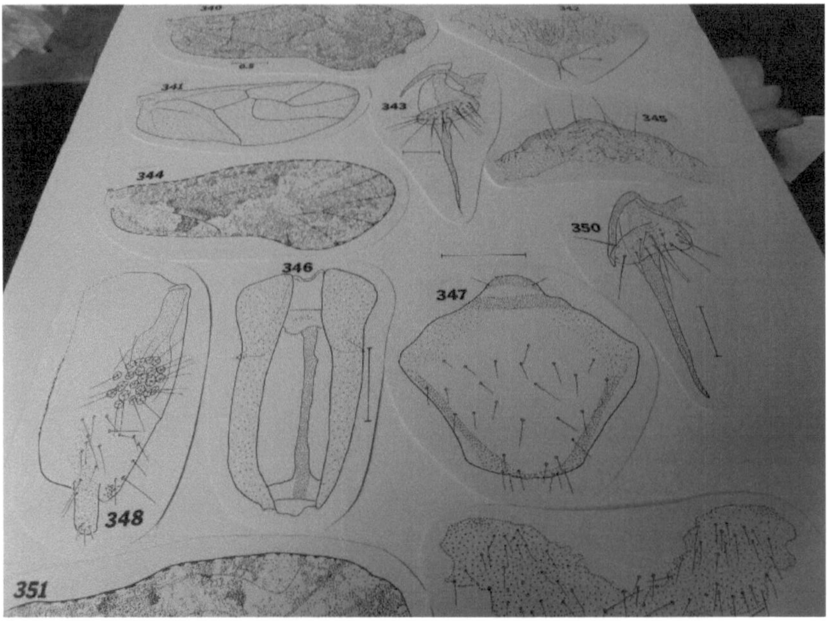

Some of Dr. Mockford's artwork of his insects (above). A photograph of another.

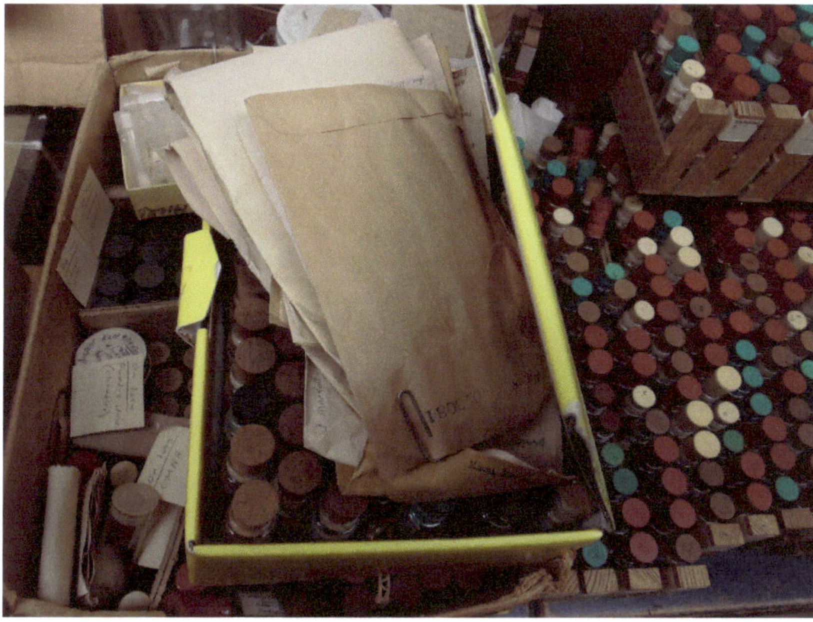

Dr. Mockford's lab was very crowded with 50 plus years of research (above). Oscar Edward Manzano exploring the lab (2014).

Multitudes of specimen vials in his lab (above). Dr. Mockford photographing a robber fly in his kitchen around 2000.

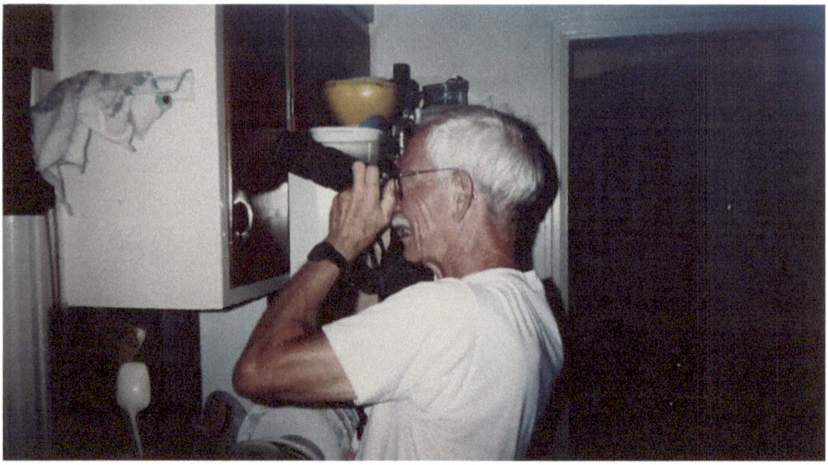

Operation Arthropod
(Operation Drop Kick)

Interview conducted on Oct 16, 2011, with Dr. Edward L. Mockford, Distinguished Professor Emeritus, concerning his research at Camp Detrick involving the use of insects as vectors for disease for warfare. Not a conspiracy theory, but fact that biological weapons were developed at Camp Detrick and were possibly used in Korea. There was propaganda at the time from the Chinese and Koreans saying so, but who would know, no? It was called propaganda by our government.

Tell us a little bit about yourself.

My name is Edward Mockford, and I'm a retired entomologist who taught entomology at the university (Illinois State University) for 26 years. I've been retired for 25 years and I still work on insects. I did my master's work at the University of Florida back in the early 50s, and I finished my master's degree in January 1954. It had been a difficult couple of years and I was pretty exhausted, and I wasn't about to start doctoral work at that time, also, you will perhaps recall that that was the period during which the Korean war was going on, and I had gotten deferments to stay out of the draft for quite some time. I think I began getting deferments while I was still an undergraduate student so, I had had quite a few deferments, and I figured they're not going to defer me forever and when they realize that I'm going to start a new program, a doctoral program, they'll probably want to draft me pretty quickly. Actually, they'd let me stay at The University of Florida for another semester and I took a couple of courses and did research on the little bugs that I was studying and at the end of that semester they drafted me.

Now, that was the end of the Korean war. It's about 1954, yeah, the shooting had stopped, but the government said we're still in the Korean emergency so the draft continued and so I got caught in it and was sent to, since I was there in Florida, I was sent to Fort Gordon, Georgia, for my basic training. After that, they sent me to a medical, forgotten what they call it, but a sort of medical supplementary unit in Savannah, Georgia, or near Savannah, Georgia, at an army installation called Fort Stewart, and I was there for a little over a month.

It became obvious to everyone involved, certainly including me, that there was nothing to do. The medical service outfits didn't do much anyway. They were supposed to inspect latrines and kitchens and it took about two people to do the inspecting for that facility. So, they decided to send me elsewhere and in about mid-August, mid to late August sometime around in there, I received orders to travel to Frederick, Maryland, which I have to admit I had never heard of at the time and to report to, they called it then, Camp Detrick, which subsequently became Fort Detrick, which was located there at Frederick. So, I went up there and I think I was flown, in fact I don't recall how I got there, but anyhow I arrived and my last transportation before arriving at the camp was by taxi, and the taxi driver clued me into what was going on at the camp . . .

I thought this kind of thing was top secret.

. . . What was happening at the camp and also you could see that there was some, I don't know, from the, I guess I had flown up there from the airport to the camp we rode by some fields which were actually part of the camp, and we could see that work was being done on plants and some of the plants looked mighty sick. In other words, they were being killed by some means that was being apparently developed at the camp, and the taxi driver assured me that yes, there were various sorts of biological experiments going on there at the camp. Well, I figured okay this sounds pretty interesting, so arriving at the camp, I think I was not assigned immediately to an area but within a few days I was assigned to a section which was run by a civilian. There were there were quite a few civilians who were working at the camp in various positions, usually administrative, and I was assigned to one run by a civilian named Dale Jenkins, and he informed me that this was a group devoted to rearing different kinds of arthropods, insects and their relatives, and we were instructed to try to develop mass rearing. Dale Jenkins informed me that that the group I had been assigned to would be involved with mass rearing, developing procedures for mass rearing of different kinds of arthropods, that is several kinds of insects and other arthropods. I got to know some of the other people involved--one of them was rearing fleas, one of them was rearing ticks, and the particular group that I was assigned to was rearing mosquitoes. Now ultimately, the idea was to develop anti-personnel weapons using these arthropods, that is we would infect them with an appropriate disease and drop a container of massive numbers of these creatures back of enemy lines, and the creatures were then to spread and spread a disease that they were carrying. That was the theory.

The purpose was to wipe out the enemy soldiers by getting them sick?

Yeah, exactly.

And what kind of diseases did you fiddle with?

We didn't get that far. We were we were primarily developing mass rearing procedures and first we started with two kinds of mosquitoes--the very well-known *aedes aegypti* which is a very good carrier, very good vector of yellow fever and can carry several other kinds of encephalitides, diseases that hit mostly the brain.

Excellent, those are my favorites.

Yeah, right, and we also well *aedes aegypti* is primarily a tropical mosquito, although it does occur in various parts of the United States at least it can get established temporarily in various parts of the United States, and it does occur regularly in the far south part of the country. But in order to cover other bases, we were also rearing a tree hole mosquito that occurs farther north and went at that time by the name *aedes triceratus.* So, this thing lives in tree hole the larvae live in tree holes ordinarily and the adults come out and bite creatures, and they are also pretty good carriers of a number of kinds of diseases, again usually encephalitides that hit the brain.

Can you give me some other of those diseases? What were some diseases that most people would know that would be in this category?

Well, nowadays this horrible West Nile Virus, but there are a number of kinds of encephalitides --Saint Louis Equine, which is primarily a disease of horses, but it will get into people. There's a whole set of them not too well known, but now and then, there'll be an outbreak of one or another of these here in the U.S. So, that's what we were up to.

It happens that *aedes* is a fairly well, *aedes* in general, is a fairly easy mosquito to rear, because they lay their eggs on whatever surface they can find right above water.

So, if you're rearing them, you have a bunch of adults in a cage, you put a container of water and some wet paper toweling in and they lay eggs on the paper toweling and then you can let them dry, and you can hold those eggs for up to several months. They'll stay alive dry up to several months and then when you want to start a batch of youngsters of larvae, you simply submerge this piece of paper with thousands of eggs on it in the water and within a day or so, you'll probably get a nice hatch of larvae. So, then you can start rearing them. Give them a little yeast, that sort of thing, and they'll get going pretty well in the water and if you want to raise large numbers you use a large tank and get the things going. That's what we were doing; we were trying to work out all the kinks involved. Now and then, I don't really know what would go wrong, but sometimes a batch of eggs wouldn't hatch, well, they'd probably been held too long or something of the sort. Now and then, larvae would sort of hold in place for a few weeks without continuing to grow, and we were trying to work out problems like that. Also, somebody else at some other lab, I don't know where, was working out procedures for developing a bomb casing for these creatures. Well, when we finally got pretty well stabilized on rearing technique with our mosquitoes, I didn't follow too closely how the guys working on the fleas, incidentally those were rat fleas that will carry plague and the people working on the ticks, I mean I knew these guys very well and they were of course, my close associates in the barracks, but we were completely involved with our own project.

The mosquitoes, incidentally, are difficult to keep in a cage; they'll ride you out--they'll get out if you put your arm in the cage to manipulate their food supply or their water supply or something or take eggs out; one or two will ride you out, and we rigged up a baffle in the room where we had the mosquitoes so that you had to pass through this sort of lock which hopefully would catch all of the mosquitoes that had ridden out on you and so not allow them to get out into the rest of the building, but still some would escape and people down the hall would complain because there are mosquitoes out and look here, we're dealing with various sorts of bacteria ourselves, and who knows what they might pick up and this was to them a serious problem. I don't think there actually was any chance of our mosquitoes picking up anything in the building, but other people weren't so convinced and in fact they got so annoyed with our mosquitoes that eventually they put us off in a Quonset hut by ourselves and let our mosquitoes go as they like, nobody else would be bothered by them.

Now this tree-hole mosquito that we were trying to rear was giving a problem in that they wouldn't mate in the cages that we had. The cages were about two feet by two feet by two feet or a little bit larger than that and these mosquitoes just wouldn't mate in there, whereas the *aedes aegypti,* the yellow fever mosquito, mated in there just fine. So, something was wrong, and my suggestion was maybe they are a crepuscular mater, that is they will only mate at a certain light intensity and so, we tried something. We tried actually, turning off the light in the rearing room and then setting up a small desk lamp and turning it away from the culture cages and observing what happened. The fact is that they started mating immediately, we just got them tumbling over each other mating in numbers and so our suggestion was that they put in a light dimmer that would bring on the light gradually in the morning and dim the light gradually in the evening. That worked just fine; we had in effect solved the problem of rearing that one. We never got it, well, actually the military guys decided right after that, well, let's not fool around with that one anymore, what that mosquito and in fact, what they wanted us to do next was try to work out a rearing procedure for the common kind of household mosquito *culex pipians* which, well, probably nobody has recognized as such, but it's a little brown ordinary sized brown mosquito, which is very common but the problem with *culex,* with any kind of *culex,* is that there is no stage that you can put aside. Here with *aedes,* we had these nice handy eggs on pieces of paper, whereas *culex* lays its eggs on the surface of water and it they have to be in contact with the water they make these little rafts, you've probably seen them on the surface of water, and they hatch right into the water and the larvae and pupae grow up in the water, and the adults then emerge out of the water. Well, this meant constantly rearing things with no possibility of holding on to any particular stage.

Another problem with them--the blood meal mosquitoes in order to well, most mosquitoes, not all, but most mosquitoes in order to produce eggs, the females have to have a blood meal. Only female mosquitos take blood, and this meant providing some means of giving a blood meal. With the *aedes* we used guinea pigs. We would anesthetize the guinea pig and put it in a wire little cylinder and put it in with the mosquitoes. Of course, they would bite the creature by the hundreds, and we'd pull it out, revive it or put it in its cage, and it would revive, scratch for a while, and it would be okay, but we were using Nembutal to anesthetize them, and that a regular dosage of Nembutal on a little animal like that is a bit too much, and they would simply die after about a couple of months of treatment like that, so we had to have new guinea pigs from time to time on that account.

Well, *culex* was another matter because *culex* prefers birds rather than mammals for its blood meal. So, we kept young chickens, kept a few young chickens around there in the lab and would Nembutalize one of those and put it in the *culex* cage and they would feed on that. It was, I thought it was just ridiculous trying to try to work with *culex* because of the many problems with it. Fortunately, we didn't have to do it, well, I didn't have to do that very long because I got out of the army before that went very far.

One aspect that was pretty interesting was the container that had been developed for use as a bomb. It was rigged up with a C02 trap. The problem is if you put a bunch of living small animals of any sort, including small insects, into an enclosed container, a relatively small, enclosed container, they're going to be producing C02, and they will gradually succumb from it. C02 is an anesthetic in strong concentrations and so you have to have some means of entrapping the C02. What you do is put in a little container of sodium hydroxide or potassium hydroxide in solution. I think that's what we did, and this will take in the C02 and change the change the salt in solution to a carbonate. The sodium associates with chloride to simply become sodium chloride, ordinary table salt and the carbon becomes, let's see, becomes a carbonate and so, gets taken out of circulation of the air and so the creatures are not going to asphyxiate themselves by C02 accumulation. Also, if you put a large number of small insects in a container, they'll tend to clump up and to avoid this there were various baffles put inside the container.

Well, we actually got around to one drop. Of course, the mosquitoes were clean; we never did, well when I was in, we never did try to infect them. We took a large number of mosquitoes down to Avon Park in Florida. There was an air force base there at the time, I don't know if it's still there or not, and we did the drop there. The drop was done on a target about twenty feet in diameter. We set out guinea pigs in their in their containers in concentric circles around the drop point. Incidentally the bomb container was rigged so that it would come open before impacting, that is when it somehow detected that it was close to the ground, it simply fell apart and the baffles containing the mosquitoes drifted down and then the mosquitoes theoretically would fly up. They hadn't been fed recently so they were expected, or many of them were expected, to fly out to the handy guinea pigs out there in concentric circles and then we were to pick them up with aspirators and so get a get an idea of the number that actually made it through and were in flying condition. What actually happened was that most of the mosquitoes, the great majority, simply sat on the baffling as though they had been stunned somehow or other and we didn't solve that problem. To this day I can't imagine really what happened except that they kind of get cowed, as it were, when placed in a container and held there for a while. Also, I still wonder if maybe the C02 trap didn't work very well, and they might have gotten a bit too much C02.

Well, that was that was pretty much the whole business. It was interesting sometimes to watch what was going on in my neighbor's projects. The fellow working on fleas had to keep rats for them to feed on and the fellow working on the ticks had to feed them on a dog. Eventually, the dog would sort of give out, which was unfortunate because since we were working in secret, we had to sort of dispose of the dog as best we could, you know, when it gave out. This was rather unfortunate.

Oh, when it came time for me to leave the army because I was only in on a two-year stint, of course, and only about nineteen months of it were to be there at Fort Detrick, my civilian boss was very anxious for me to stay on. I told him, not a chance. Oh, he said, I'll get you a sergeant's rating, you'll just have a great time. Well, I didn't enjoy it all that much and I told him, look here, I'm I have applied to graduate school at two midwestern universities for doctoral work and I'm determined to get on with it as quickly as possible, and I'm not going to stay an extra day in the military and that's all there was to that.

The barracks life there, although barracks life was never very pleasant, was I would say extremely interesting in that almost all of the troops had advanced degrees. Some of them, especially the engineers, had only bachelor's degrees, they thought they were big stuff, but most of us had master's degrees and there were one or two PhDs among the draftees. One of my one of my good friends there, Dr. Mitchell Byrd, had his doctorate before he was drafted, and I gather there were one or two others who also had doctorates. There were many of us were planning to get back into university and do doctorates and one of my very good friends who was working on the tick project—

How many of these insect projects were there?

Ticks, fleas, and mosquitos. There were all sorts of other projects that went on there on various sorts of bacteria and all sorts of horrible creatures. I didn't know much about those, but I would hear about them now and then.

What do you mean by horrible creatures, other viruses?

Yeah, brucellosis was being . . .

What is brucellosis?

It used to be before dairy products were always pasteurized, this disease would be carried in the milk, and it results in a disease called undulate fever which is pretty bad. It just debilitates a person and then they were working on anthrax. Also, one aspect of life at Fort Detrick was that all the troops had to get shots for all sorts of diseases So, I got shots for plague and all sorts of things that normally you just wouldn't think of.

I think I'm going to go down to Walgreens and get my plague shot.

The mosquito fella named David North, who had a master's degree from Penn State, I guess it was, was needle shy and he would pass out whenever they took him for the set of vaccinations. Fortunately, I wasn't---I didn't care if they filled me full of this stuff, it was fine, but that was life at Fort Detrick.

By the way, in the army in those days, the troops still did K.P. (kitchen patrol) so now and then, I would get pulled out of the lab and help clean up the kitchen, which was kind of interesting. Another aspect of it was kind of nice; we were only about thirty-five miles from Washington D.C., north of Washington D.C. and if your work was pretty much caught up and you had nothing else to do, you could sign out, go down there to the U.S. Department of Agriculture Library and do library work on your project, which was rather nice. I got to know the mosquito literature at the time quite well by that means. Also, of course being close to Washington, we could get leave on weekends to go down there, and so I got to work on my own creatures at the U.S. National Museum. I got a pass to go in there on weekends and work in the research area on my own bugs, so that was rather nice. So, that was life at Camp Dietrick.

. . . While working on a biological weapon.

Yeah, it was great fun.

And so, the project, you never had a name for it other than you called it Project Arthropod?

Yes, I guess that was it.

That's interesting. Well, that's always been a big conspiracy theory. There's a lot of people who have said that a lot of weird stuff has gone on at Fort Detrick. Well, thank you for sharing that with us.

It was an interesting nineteen months of my life.

Thank you, doctor.

I have a *Padrino*, a godfather, not one of those gangster clichés—ain't gonna break no legs, he's a professor, studies *psocids*—book lice, an expert in the field, many species under his belt including world's smallest insect—a wasp that lays eggs on the eggs of *psocids* and *mi padrino* Mockford has spent his life studying these creatures and it all began for him on an Indiana summer night with cicadas singing and, on the day, when the colors on the wings of butterflies captivated his soul and somehow, he ended up my Godfather, maybe due to the fates being *psocids* and he got me into collecting when I was like 7—that ended when I was like 8—saw this black-purple-bruise-like-beetle thought it would be perfect for my collection, the majority of which was eaten by other bugs who eat crunchy dried bug corpses so that eventually the cork lined box ended up as a storage box for my father's fighting cock knives—he had some from the Philippines called razors and little boxing gloves and gaffs which are like sharp needles, all for cockfighting, and my father is actually one of the most compassionate-towards-animals-people I know, and this beetle was crawling along—I couldn't kill it—I knew I could snag it easily and freeze it and pin it in the box with the other husks, just couldn't though—felt bad to even want to kill it—but I still have my copy of *Borror's and White's Field Guide to the Insects* on my bookshelf, which my Godfather had given me before the beetle incident--where I couldn't do it in—watching it shambling along in the leaves was enough—glorious colors of oil on water mixing on its black back as it did its beetle life and I mine, and my Godfather his, which sometimes involves the death or stunning (with Alka Seltzer) of a *psocid*, but my Godfather is one of the most compassionate people I know regarding life of any kind—and I may have got that feeling from either God or Father more than likely it seems to me Now, both_

First edition June 6, 2023

www.ingramcontent.com/pod-product-compliance
Lightning Source LLC
Chambersburg PA
CBHW041113180526
45172CB00001B/228
9781312483064